巨藻之舞

The Kelp Dance

Gunter Pauli

冈特·鲍利 著

郭光普 译

丛书编委会

主　任：贾　峰

副主任：何家振　郑立明

委　员：牛玲娟　李原原　李曙东　吴建民　彭　勇
　　　　冯　缨　靳增江

丛书出版委员会

主　任：段学俭

副主任：匡志强　张　蓉

成　员：叶　刚　李晓梅　魏　来　徐雅清　田振军
　　　　蔡雩奇

特别感谢以下热心人士对译稿润色工作的支持：

姜竹青　韩　笑　杨　爽　周依奇　于　哲　阳平坚

李雪红　汪　楠　单　威　查振旺　李海红　姚爱静

朱　国　彭　江　于洪英　隋淑光　严　岷

目录

巨藻之舞	4
你知道吗?	22
想一想	26
自己动手!	27
学科知识	28
情感智慧	29
艺术	29
思维拓展	30
动手能力	30
故事灵感来自	31

Contents

The Kelp Dance	4
Did you know?	22
Think about it	26
Do it yourself!	27
Academic Knowledge	28
Emotional Intelligence	29
The Arts	29
Systems: Making the Connections	30
Capacity to Implement	30
This fable is inspired by	31

一棵庞大的巨藻坐在纳米比亚海岸边清凉的海水里,随着海浪摆动着。几头海狮穿梭在这些大海藻间,寻找着鱼吃。
"你真是一棵神奇的植物,"一头海狮有点害羞地赞美道,"你真的和竹子长得一样快吗?"

A giant kelp sits in the cold waters off the Namibian coast, moving with the waves. Some sea lions are swimming swiftly through the kelp forest, looking for fish.

"You are an amazing plant," remarks a sea lion shyly. "Is it true you grow as fast as bamboo?"

你真是一棵神奇的植物

You are an amazing plant

你能长多高呢?

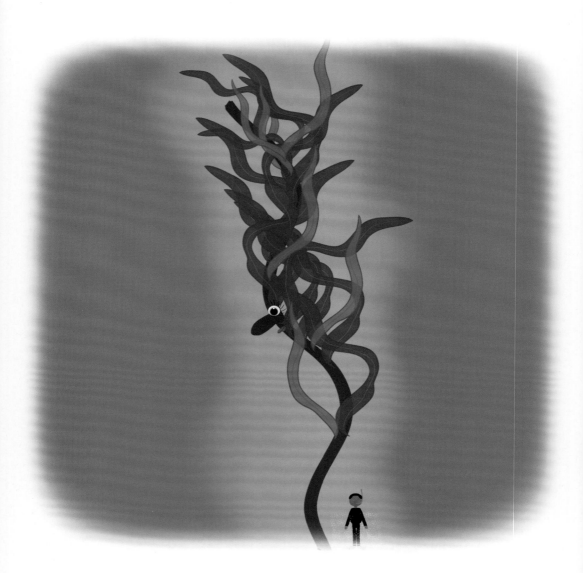

How tall will you grow?

"是的,我每天能长50厘米。"巨藻回答。

"你能长多高呢?"

"噢,我现在还是个少年,不过只要人们不过度捕捞,我可以长到50米高。"

"You are right – I can grow 50 centimetres per day," answers the kelp.

"And how tall will you grow?"

"Well, I'm still a teenager but unless people overfish, I can grow 50 metres tall."

"这可比最高的竹子还高一倍呢!"海狮回答道,"太神奇了!可你能告诉我过度捕捞与你长得高大健壮有什么关系呢?"

"哎!人类喜欢吃鱼,而鱼的食物又吃巨藻。所以如果人类捕捞太多的鱼,那些喜欢吃我们的家伙就会大吃特吃,我就长不大了。"

"That is double the size of the biggest bamboos!" replies the sea lion. "That's amazing. But tell me, what does overfishing have to do with you growing tall and healthy?"

"Well, humans like to eat fish, and fish eat stuff that eats kelp. So if people catch too many fish, then those who like to eat me will eat so much that I cannot grow."

……那些喜欢吃我们的家伙就会大吃特吃……

...those who like to eat me will eat so much...

我们是海洋里的森林

We are the forests of the sea

"这么说,鱼群越多、越活跃,巨藻就长得越好。这是一个真实的生态系统。"海狮说。

"我们能长得很大。我们是海洋里的森林,为你们和海獭提供了丰富的食物。"

"So, the more fish are alive and well, the more the kelp will grow. That is a real ecosystem," says the sea lion.

"We grow big. We are the forests of the sea, providing an abundance of food for you and the sea otters!"

"我知道你和海獭是好朋友。"

"我们沿着太平洋组成了巨型森林,帮助早期的航海者家航行。我们是海洋中的巨藻高速公路,指引人们在长距离航行时也能得到充足的食物。"

"人们是沿着巨藻高速公路旅行的吗?"

"当然,等到达彼岸他们就可以和之前一样生活。只要有巨藻的地方就有龙虾、石斑鱼和鲍鱼。我们甚至还能产生能量!"

"I know you and the sea otter are dear friends."

"Our huge forest along the Pacific Ocean helped early seafarers to navigate. We were the kelp highway of the ocean, guiding people over long distances with sure supplies of food."

"People travelled following the kelp freeway?"

"Sure, and when they got to the other side they could live on like before. Anywhere you have kelp, you also have lobsters, rockfish and abalone. We can even produce energy!"

……我们沿着太平洋组成了巨型森林……

...our huge forest along the Pacific Ocean...

...酒精和面条...

...ethanol and spaghetti...

"巨藻能制造食物和燃料？"

"是的，我们富含矿物质和碘。如果你把我们收集在一起并放到池子里发酵，最后你就会得到大量沼气。你还能用我们制糖，或生产酒精给汽车作燃料。有些聪明的人们还把我们做成意大利面条食用。"

"你们能生产这么多好东西！为什么我们对你们了解这么少？"海狮问道。

"Kelp makes food and fuel?"

"Yes, we are full of minerals and iodine. If you harvest me and let me rot in a closed room, you will end up with plenty of biogas. You can make sugars from me and produce ethanol to fuel a car. Some smart people turn me into spaghetti and eat me."

"You make so many goodies. So why we know so little about you?" asks the sea lion.

"你知道我们有
自己的舞蹈吗,海狮?"
"舞蹈?噢,是的,我听说你们
好像还启发了跳肚皮舞的人!"
"我们随着海洋的节奏跳舞。"

"Did you know we have our own dance, Sea Lion?"
"Dance? Well, yes, you seem to have inspired the belly dancers, from what I've heard!"
"We dance with the rhythm of the sea."

……启发了跳肚皮舞的人……

...inspired the belly dancers...

用海啸的力量跳舞

Dance with the power of the tsunami

"那要是海啸来了怎么办呢?"
"我们会用海啸的力量跳舞!"
"那怎么做?"

"And, what if there is a tsunami?"
"We will dance with the power of the tsunami!"
"How is that?"

"当海洋的力量太强时,我们都匍匐在海底,任由海水翻涌而过。"

"然后你们就又可以继续跳肚皮舞了,就像歌星莎吉拉教我们的一样!"

……这仅仅是开始!……

"When the power of the ocean is too strong, we all lie flat with our faces on the ground, and let it just flow over us."

"And then you can belly dance away again, just like the singer Shakira has taught us!"

... AND IT HAS ONLY JUST BEGUN!...

……这仅仅是开始!……

... AND IT HAS ONLY JUST BEGUN! ...

Did You Know?

你知道吗?

Kelp can grow 50 metres tall. Its stem is flexible, allowing it to sway in any direction following the ocean currents.

巨藻能长到50米高。它的茎杆是有弹性的，能随着洋流向任何方向摇摆。

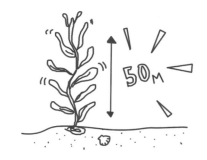

Kelp roots do not carry nutrients to the leaves. They are only an anchor. The stem is called a frond and the leaves are called blades.

巨藻的根不会为叶子供应营养，它只是一个锚。巨藻的茎杆称为藻体，而叶子称为叶片。

Kelp is a popular animal feed. The leaves can photosynthesise on both sides. (The leaves on trees do it on one side only.) Kelp leaves drop off once a month, providing lots of feed.

巨藻是一种常见的动物饲料。巨藻的叶子两面都可以进行光合作用。（陆地上大树的叶子只能一面进行光合作用。）巨藻叶片每个月脱落一次，可以提供大量饲料。

Kelp forests are extremely rich and diverse, supporting an abundance of shellfish, marine mammals, seabirds and seaweeds. There are kelp forests along both the Pacific Rim and Southern Africa.

巨藻森林具有极大的丰富性和多样性，支持着大量的贝类、海洋哺乳动物、海鸟和海藻的生活。环太平洋和南非都有巨藻森林。

Kelp forests reduce wave power, offer grip and support for boats, and create an abundance of food. These forests may have facilitated the emigration of people from Asia to the Americas over the North Pacific Coast. The kelp ecosystem is therefore known as the "kelp highway".

巨藻森林减弱了海浪的力量，为船只提供固定物和支持，还能提供大量食物。这些森林很可能帮助过亚洲人穿过太平洋北部海岸迁徙到美洲。巨藻生态系统因此被称为"巨藻高速公路"。

The giant bamboo (Guadua angustifolia) is only half the size of the giant kelp (Macrocystis pyrifera). Giant bamboo can grow up to 90cm per day, and giant kelp up to 50cm. Bamboo dies when it flowers, and lives up to 70 years; kelp can live up to 100 years.

一种叫瓜多竹的巨竹也只有巨藻的一半高。巨竹每天能生长90厘米，而巨藻每天只长50厘米。竹子开花后就会枯死，可以活70年，而巨藻能活一百多年。

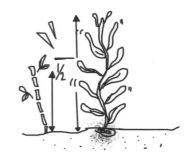

Kelp has the highest concentration of iodine of any food. Kelp noodles are made from kelp, sodium alginate and water. It is free from gluten, fat, cholesterol, protein and sugar.

在所有食物中，巨藻含碘量最高。巨藻面条就是用巨藻、藻酸钠和水做成的，不含面筋、脂类、胆固醇、蛋白质和糖。

Giant kelp is called the Sequoia of the Seas. It moves graciously with the flow of the water as if dancing.

巨藻被称为海洋中的红杉。它随着海水优雅地摆动，就像是在跳舞。

Think About It 想一想

Kelp relies on a healthy population of fish to thrive, but does safeguarding fish populations necessarily translate into more kelp?

巨藻依赖健康的鱼类种群而茁壮生长，但是鱼群越多是不是就必然能生成更多的巨藻呢？

Is it important to be the greatest, the fastest, the oldest and/or the best?

成为最大、最快、最老和／或最好真的很重要吗？

当面对一股势不可挡的强大力量时，你会怎么办呢？

What would you do if faced with an overwhelming force you could not stop?

Which music inspires you to dance?

什么音乐会让你跳起舞来？

Do It Yourself! 自己动手!

Go to your local shops. First pass by a supermarket and ask if there are any seaweeds for sale. Most likely the answer will be "No". Then ask which products contain seaweed. Most likely you will get an "I don't know" answer. The answers you might receive in Japan will be different from those received in France (Brittany) or Spain (Galicia). Check ice cream, microwave meals and frozen-food labels yourself. You will likely recognise the word "carrageenan", or perhaps a general description like "emulsifier", but you might not have known that these are made from seaweed. Now stop at the local health-food store. Ask if there are any seaweeds for sale or if it is included in any of the products. You are likely to be listening to the answer for the next 10 minutes. Seaweed could be everywhere.

去当地的商店看看。先去超市问问是否有海藻卖，答案很可能是"没有"。然后问下什么产品含有海藻，最可能的答案是"我不知道"。你在日本得到的答案可能与在法国布列塔尼和西班牙加利西亚的不同。检查一下奶油、微波炉烤肉和冷冻食物标签，你可能会发现"角叉菜胶"的字样，或者"乳化剂"之类的一般性描述，但是你可能不知道这些东西都是从海藻中提取出来的。现在到当地的健康食品商店看看，问问是否有海藻卖，或者是不是有什么产品含有海藻。你很可能会听到10分钟都说不完的答案。海藻无处不在。

TEACHER AND PARENT GUIDE

学科知识
Academic Knowledge

生物学	巨藻富含碘，碘会促进甲状腺激素的合成，调节食物代谢，同时也解毒；自闭症与缺碘有关；营养、能量和生长是按照生态系统中的一套制衡机制运行的；巨藻是一种褐色水藻。
化 学	用海藻做的纤维能用来缝合伤口；海藻中的钙离子和体液中的钠离子可以使氧气渗透到细胞里，促进组织生长。
物 理	冲击和断续的运动，正弦运动；藻酸钠可增强黏度并用作乳化剂；巨藻的浮力使它们克服重力而漂浮在水中。
工程学	与其对抗地球和海洋的运动，不如随着这些节奏一起运动，这是针对自然力量的最好防御方式；海藻纤维做的毛巾已经于1993年开始供应市场；海藻纤维泳装于2000年出现，2012年又出现了淡藻纤维做的婴儿服装。
经济学	一吨干的巨藻能生产200～250千克海藻酸盐，可以用来制成200千克纤维；用巨藻生产的产品有100多种。
伦理学	如果一个生态系统内所有组成部分的功能都能发挥到最好，这个生态系统就能繁荣发展，所以一种生物不能过度开发利用另一种生物；生态系统需要地球上的所有成员相互协调，以达到一个动态平衡。
历 史	舞蹈作为一种宗教仪式出现在古代埃及和希腊。
地 理	大型巨藻森林位于太平洋中；巨藻森林沿着非洲海岸分布，从南非到纳米比亚，沿途都受本格拉洋流的影响。
数 学	浮力的计算；海啸波的数学特征。
生活方式	海白菜和裙带菜是最常见的食用褐藻；舞蹈是一种情绪表达方式，也是一种治疗方法。
社会学	肚皮舞和西班牙的弗莱明戈舞很相似，起源于宗教舞蹈，是在有关生育的仪式上和与孩子有关的仪式上演出的。
心理学	即使你觉得你是最好的，承认某个人或某件事可能在某些方面比你好也是一种谦逊的态度。
系统论	我们太依赖于土地上的产品了，我们唯一利用的海洋资源就是鱼，而在海藻森林中却蕴含着巨大的未开发的生物量，只要我们控制捕鱼量，就能让这些生物量可持续地提供一百多种产品。

教师与家长指南

情感智慧
Emotional Intelligence

海狮

海狮很羡慕巨藻，并害羞地赞美她。她想了解更多，主动提出了很多问题。她很专心，学得很快，理解了有关生态系统新的信息及其重要性。随着理解加深，海狮认识到还有很多知识她还不了解。她听到了新的理论（巨藻高速公路），这些一鳞半爪的信息呈现给她的东西比她之前想象的还要多。巨藻有如此的节奏和灵活性，还知道如何应付海啸，这让海狮感到惊奇。最后，海狮赞美了巨藻，并谈到了当代最著名的音乐家和肚皮舞舞蹈家之一的莎吉拉，以此表达了对巨藻的仰慕。

巨藻

巨藻很镇静，但很了解自己的能力、时代和位置。巨藻很享受海狮的陪伴，花时间回应和解答她的困惑，包括对其他因素（鱼）的依赖和自己无法控制的威胁（过度捕捞）。巨藻还顺便提到了自己的重要历史价值。巨藻没有狂妄自大，恰如其分地分享了她的潜力和才能。然而，巨藻很有自知之明，并意识到了自己可能遇到的事情和自己的局限性。她的舞蹈让海狮大吃一惊，她还告诉海狮自己应对海啸的神秘能力，这也是她们的友谊中互相信任的表现。

艺术
The Arts

该上舞蹈课了！让我们用阿梅利亚·特拉平（AMELIA TERRAPIN）来编排的舞蹈，在运动中学习科学。怎么做呢？莫比乌斯网站（MOBIUSMOVES.COM）介绍说你需要学习基本的舞蹈动作。然后和朋友们讨论：当海啸来的时候你们怎么跳舞？在小组面前试一下，一起商量出一种最好的方式通过舞蹈来表现"海啸"。现在尝试表演做生物饮料以及和鱼做朋友。每次都要想一下怎么用舞蹈动作来"说话"，而不能真的说话。阿梅利亚做这个已经很多年了，所以她是个伟大的教练。

TEACHER AND PARENT GUIDE

思维拓展
Systems: Making the Connections

巨藻是一种卓越的高产生态系统:很少有人深刻认识到它是世界上生长最快、用途最多的生物之一。巨藻不仅在海洋森林生态系统中是个关键物种,还是可持续性产业的基础。它应用于医疗、土壤肥力、食品、纺织和能源等数百种产品中;而且我们只要收获就行,从来不用种植它们。你能想象出如果庄稼只需收获而不用种植,也不用管理,这会节约多少钱和时间吗?巨藻提供的纺织材料可以用咸水加工处理,比起棉花节约了很多淡水。但是棉花纤维只能来源于种子,而巨藻中的海藻酸盐占整个植株生物量的25%。尽管可以从大多数巨藻(和其他海藻)中获取的产品都有明确的记载,并在科学上得到证明,甚至在世界范围都有生产,然而没有一种产品成为主流产品被使用。旅游业创造了新的探险活动,比如潜水到有防护的大白鲨洞穴中;但是,还没有旅游者去观看同样精彩壮观的巨藻森林,以及其中的鱼群和海獭。主要原因可能是无知:人们不知道巨藻森林里有如此多姿多彩的环境。由于过度捕鱼,巨藻森林正在急剧消失,甚至红树林也不得不为养殖业(主要是养虾)和旅游腾出空间。我们不仅仅要知道巨藻控制海浪和应对最猛烈的海啸的生存能力,还要知道它保护着海洋森林中的所有生命。人们还不了解巨藻对生态系统的这些贡献:未来的岁月会证明巨藻森林和热带雨林一样重要。

动手能力
Capacity to Implement

海藻能为未来提供很多食物、医药、纺织品、肥料和清洁产品,随着需求增长,供应也将越来越快。在你生活的地方找一下有谁为工业生产提供海藻原料。如果现在还没有,想象一下你自己变身海藻产业背后的推动力。你会注重什么?哪种产品最有意义?开始你会做原材料的供应商吗,或者促进现有产品的销售?想象一下你如何把海藻产业变成你的谋生之道。

教师与家长指南

故事灵感来自

莎吉拉·伊莎贝尔·梅巴拉克·里波利
Shakira Isabel Mebarak Ripoll

莎吉拉出生在哥伦比亚西北部的港口城市巴兰基利亚，并在那里长大。她父母分别是哥伦比亚人和黎巴嫩人。她在学校的时候就开始表演。莎吉拉把肚皮舞再度介绍给更多的观众并鼓舞了数百万人想和她一样跳舞。她一直是销量最好的艺术家之一，她的唱片在全世界销售了1300多万张。她在2010年国际足联世界杯的主题歌曲《非洲时刻》是历来最受欢迎的音乐视频之一。莎吉拉通过赤足基金会建立了多所有舞蹈演播室的学校，以帮助哥伦比亚的贫困孩子。

更多资讯

www.tandfonline.com/doi/abs/10.1080/15564890701628612#.VLFKgCHqwA

textilemachine.fangzhi-jixie.com/machine_1/content/?1110.html

seaweedindustry.com/seaweed/glossary

mobiusmoves.com/

pbskids.org/dragonflytv/show/kelpforest.html

图书在版编目（CIP）数据

巨藻之舞：汉英对照 /（比）鲍利著；郭光普译. —— 上海：学林出版社，2015.6
（冈特生态童书. 第2辑）
ISBN 978-7-5486-0872-1

Ⅰ. ①巨… Ⅱ. ①鲍… ②郭… Ⅲ. ①生态环境－环境保护－儿童读物－汉、英 Ⅳ. ① X171.1-49

中国版本图书馆 CIP 数据核字 (2015) 第 092526 号

————————————————————————————

ⓒ 2015 Gunter Pauli
著作权合同登记号 图字 09-2015-446 号

冈特生态童书
巨藻之舞

作　　者——	冈特·鲍利
译　　者——	郭光普
策　　划——	匡志强
责任编辑——	李晓梅
装帧设计——	魏　来
出　　版——	上海世纪出版股份有限公司 学林出版社
	地　址：上海钦州南路81号　电话/传真：021-64515005
	网　址：www.xuelinpress.com
发　　行——	上海世纪出版股份有限公司发行中心
	（上海福建中路193号 网址：www.ewen.co）
印　　刷——	上海图宇印刷有限公司
开　　本——	710×1020　1/16
印　　张——	2
字　　数——	5万
版　　次——	2015年6月第1版
	2015年6月第1次印刷
书　　号——	ISBN 978-7-5486-0872-1/G·321
定　　价——	10.00元

（如发生印刷、装订质量问题，读者可向工厂调换）